NEXTGEN LEARNERS PRESENTS

Think Like a Brain!™ Collection

Brainstem Benny & The Automatic Adventure

Keeping Your Body Running Without Even Thinking

BOOK 2

Think Like a Brain!™
Collection Overview

Think Like a Brain!™ is a wildly engaging, silly-smart children's book collection that invites kids to explore how their brains work from the inside out. Through playful characters, imaginative settings, and rhythm-rich storytelling, this collection teaches neuroscience, social-emotional learning (SEL), and self-understanding in ways children absorb without even realizing they are learning.

Instead of telling children how to behave or what to feel, Think Like a Brain!™ shows them why their brains respond the way they do. Each story transforms complex brain science into friendly characters and relatable moments, helping kids feel capable, curious, and empowered rather than confused or ashamed.

Designed for children first and embraced by parents and educators as preventative education, this collection supports:

- Emotional regulation and self-awareness
- Executive functioning skills
- Neurodivergent understanding and inclusion
- Brain-body connection and growth
- Compassion for self and others

With humor, heart, and factual accuracy at its core, Think Like a Brain!™ helps children build the skills many adults wish they had learned sooner, one playful page at a time.

One of the collection's most powerful features is how the characters become real-life reminders for children long after the story ends. Once kids learn who the characters are and what jobs they do, they can begin to recognize those same processes in themselves. A big feeling might become, "It looks like Amy is trying to tell you something," or a moment of pause might sound like, "What does your Fred think?" By naming experiences through familiar characters, children learn to notice what is happening inside their brains, reflect on it, and move forward more quickly with understanding rather than frustration.

Series 1:
The Brainy Bunch (Brain Anatomy)

Series Overview: Meet the brain's core parts as a friendly team. Kids learn what each part does and how they work together to help thinking, feeling, moving, and learning.

1. **The Brainy Bunch**
   Focus: Brain overview and teamwork
2. **Brainstem Benny & The Automatic Adventure**
   Focus: Breathing, heartbeat, survival basics
3. **Bella Cerebellum's Balance Ballet**
   Focus: Balance, coordination, practice
4. **Olly Occipital & The Lookout Tower**
   Focus: Vision, visual processing
5. **Tammy Temporal & The Soundy Sound Storm**
   Focus: Hearing, language, sound processing
6. **Perry the Parietal & The Body Map Bash**
   Focus: Touch, space, body awareness
7. **Hippocampus Hal & The Memory Maze**
   Focus: Memory, learning, recall
8. **Amy the Amygdala & The Overactive Alarm**
   Focus: Emotional alarm system
9. **Luna Lefty & Roger Righty's Mixed Up Masterpiece**
   Focus: Left/right hemispheres, teamwork
10. **Callie the Connector & The Brain Bridge**
    Focus: Corpus callosum, brain-to-brain communication

Series 2:
Cortex Castle (Frontal Lobe Executive Functions)
Series Overview: A lively castle of executive-function helpers. Kids practice planning, starting, stopping, focusing, flexible thinking, time skills, and self-talk, with Frontal Fred guiding thoughtful choices.

1. **Cortex Castle**
   Focus: Executive function overview
2. **Frontal Fred & The Decision Dilemma**
   Focus: Decision-making, prioritizing
3. **Impulsive Ian & The Button Box**
   Focus: Impulse control, pause
4. **Memory Molly & The Wandering Thoughts**
   Focus: Working memory, holding steps
5. **Focus Frannie & The Great Distraction Chase**
   Focus: Attention, refocusing
6. **Planella & The Checklist Parade**
   Focus: Planning, sequencing
7. **Flexi & The Changey-Changeroo Day**
   Focus: Cognitive flexibility
8. **Self-Talk Sally & The Thoughtful Echoes**
   Focus: Self-talk, emotional regulation
9. **Starter Stacy & Finisher Finn**
   Focus: Task initiation, task completion
10. M**ilo the Minute Keeper & Mae the Moment Seeker**
    Focus: Time awareness, pacing
11. **The Power of Pausing with Penny**
    Focus: Pause skills, regulation

Series 3:
Busy Brain Café (Neurodivergent Affirmations)

Series Overview: A welcoming café where different brain styles are celebrated. These stories affirm neurodivergent experiences while teaching self-understanding, communication, and compassion for others.

1. **Nathan Neuron & The Bouncy Brain**
   Focus: ADHD metaphor, movement, attention
2. **Sara Sprinkle the Sensa & The Super Sensitive Day**
   Focus: Sensory sensitivity, overload tools
3. **Overwhelmed Owen & The Senses Overload**
   Focus: Sensory overload, coping strategies
4. **Interoceptive Ivy & The Inside Signals**
   Focus: Interoception, body signals
5. **Zane's Zigzag Thoughts & The Path to Peace**
   Focus: Anxiety, grounding
6. **The Busy Brain Café**
   Focus: Neurodivergent inclusion, different processing styles
7. **The Listening Light**
   Focus: Nonverbal communication, understanding
8. **The Language of Love**
   Focus: Connection, communication, compassion
9. **Trying Theo & The How-To Hiccup**
   Focus: Persistence, frustration tolerance
10. **Drowsy Drew & The Nighttime Repair**
    Focus: Sleep, brain repair
11. **Burned-Out Bonnie & The Rebuild**
    Focus: Burnout, rest, recovery

Series 4:
Steady Systems Squad (Regulation & Growth)

Series Overview: A squad that explains how brains regulate, reset, rest, and grow stronger over time, including stress waves, sleep crews, sensory filtering, and rebuilding after hard days.

1. **The Steady Systems Squad**
   Focus: Regulation systems overview
2. **Timothy Thalamus & The Traffic Tower**
   Focus: Sensory filtering, attention routing
3. **Danny Dimmer & Wake-Up Wonda**
   Focus: Arousal and alertness
4. **The Start, Stop, and Switch Station**
   Focus: Initiation, inhibition, shifting
5. **The Night Shift & The Sleepy Crew**
   Focus: Sleep systems, restoration
6. **The Stress Wave & The Return to Calm**
   Focus: Stress response, recovery
7. **The Brain Growth Gathering**
   Focus: Brain growth and development
8. **Paul's Practice Path**
   Focus: Neuroplasticity
9. **Daisy's Daydream Network**
   Focus: Daydream network, imagination
10. **Remy's Rebuild Crew**
    Focus: Recovery and healing

NEXTGEN LEARNERS PRESENTS

Think Like a Brain!™ Collection

Brainstem Benny & The Automatic Adventure

Keeping Your Body Running Without Even Thinking

BOOK 2

WRITTEN &
ILLUSTRATED BY:
MAISEL MCLAULA

NEXTGEN LEARNERS PRESS™
CINCINNATI, OHIO

This book is published by NextGen Learners Press, an imprint of NextGen Learners LLC.

Maisel McLaula is a pen name used by the author of this work.

First Edition: 2026
Printed in the United States of America
ISBN: 978-1-972860-01-4

This book is intended for educational and informational purposes only. It is not intended to replace professional medical, psychological, or therapeutic advice, diagnosis, or treatment.
Always seek the advice of a qualified professional with any questions regarding a medical or mental health condition.

Artwork and visual elements in this book were created using a combination of original artwork, with enhanced original direction, digital illustration tools, and AI-assisted creative technologies.

A Note for My Readers:

Hi friend,
Your brain is incredible. It thinks, feels, remembers, solves, dances, creates, and so much more!
This series is here to help you understand how your brain works, and to cheer you on as you grow your own thinking superpowers.

Every part of your brain has a job.
Sometimes they get mixed up. That's okay!

When you learn what each part does, you can work with your brain instead of feeling like it's working against you.
So read, giggle, learn, and maybe even draw your own Brainy Bunch when you're done.

I'm cheering you on with every rhyme.

-Maisel McLaula

Think Like a Brain!™

Brain Anatomy Collection

Explore the castle inside your head! Meet each room and its brilliant resident, from Frontal Fred to Benny the Brainstem. Whimsy, wonder, and neurons await!

For Cayleen, our Sheriff.
You showed up with quiet care,
helping things run smoothly without being asked.
Like Benny, you did your part in steady,
thoughtful ways that made a difference.

And...

To the beautifully wired minds whose brains keep working quietly to keep them safe.

You are exactly what a great brain looks like.

- Maisel McLaula

In the castle's basement where it's quiet and neat,
Lived Benny the Brainstem, calm and discreet.
He kept things running from morning to night,
Like breathing and blinking without any fright.

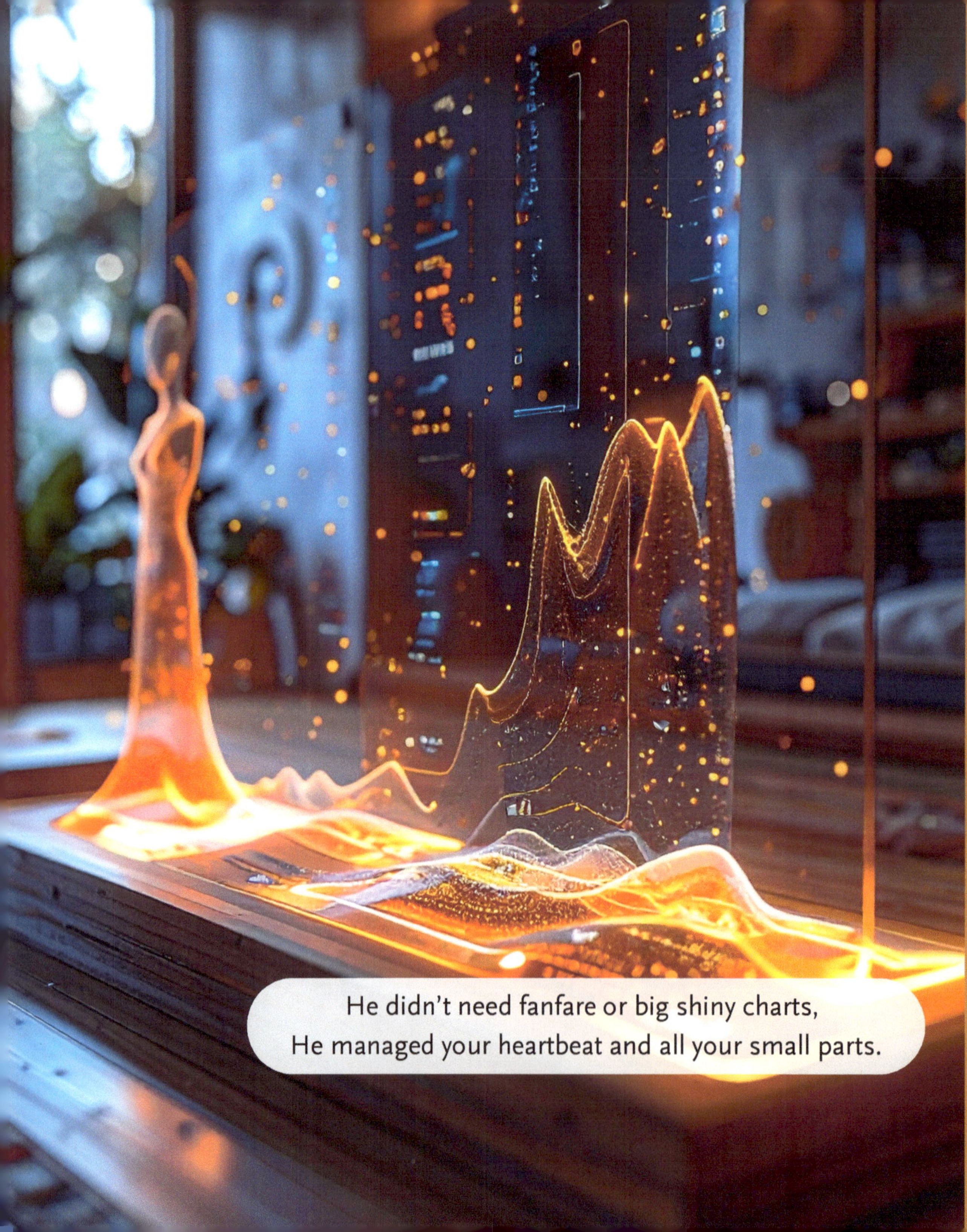

He didn't need fanfare or big shiny charts,
He managed your heartbeat and all your small parts.

He kept you digesting and standing up tall,
And even helped wake you when morning birds call.

He clicked on your blinking and managed your yawn,
And kept your legs moving from dusk until dawn.

He whispered, “Stand straight,” and “Don’t forget air!”
But most folks forgot he was even there.

One day, the castle grew dizzy and slow,
Fred's notes were flying and no one could go.

"Why do I feel woozy?" cried Bella mid-bow.
And Perry fell backwards and shouted, "Wow!"

Tammy was tangled, Olly saw blur,
Amy's alarms went "Whirr! Whirr! Whirr!"

Frannie lost focus, Hal forgot math,
The Brainy Bunch stumbled down Benny's old path.

They knocked on his boiler room door in a spin,
"Benny, we're scrambled! Please, let us in!"

Benny just blinked and turned down the light,
"Let's sit and breathe. You'll start to feel right."

He showed them deep breathing and rest that's secure,
He said, "I'm the base, you need me for sure.

I'm not very flashy, I'm calm and I'm still,
But without my work, you'd lose all your skill."

He taught them to stretch and to pause with a hum,
To notice their heartbeat, their breath, and their thumbs.

He helped them feel grounded, not scattered or stuck,
And slowly their balance returned with some luck.

Their eyes became clearer, their toes tapped in time,
Their breathing slowed down and their rhythm aligned.

Fred whispered, “Dear Benny, we forgot your role,
But your work is the center, the base, and the goal!”

Benny just nodded and blinked once again,
"It's okay, dear friends, I'm always within.

I’m here when you sleep, and I’m here when you run,
I work in the dark, but I shine like the sun.”

They built him a badge that shimmered and glowed,
With “Boiler Room Boss” in glittery code.

Benny just chuckled and straightened his vest,
He was happy to help and proud of his best.

So if you feel dizzy or tired or slow,
And your energy's dropping or movements won't flow,

Take time to breathe, to sip, and to rest,
Your brainstem is working, and doing its best.

Benny's Boiler Room Tips

- Practice deep breathing to reset your state
- Drink water often and go at your rate
- Stretch every morning, then stretch again slow
- Pause in the quiet and let your breath flow

What Is the Brainstem?

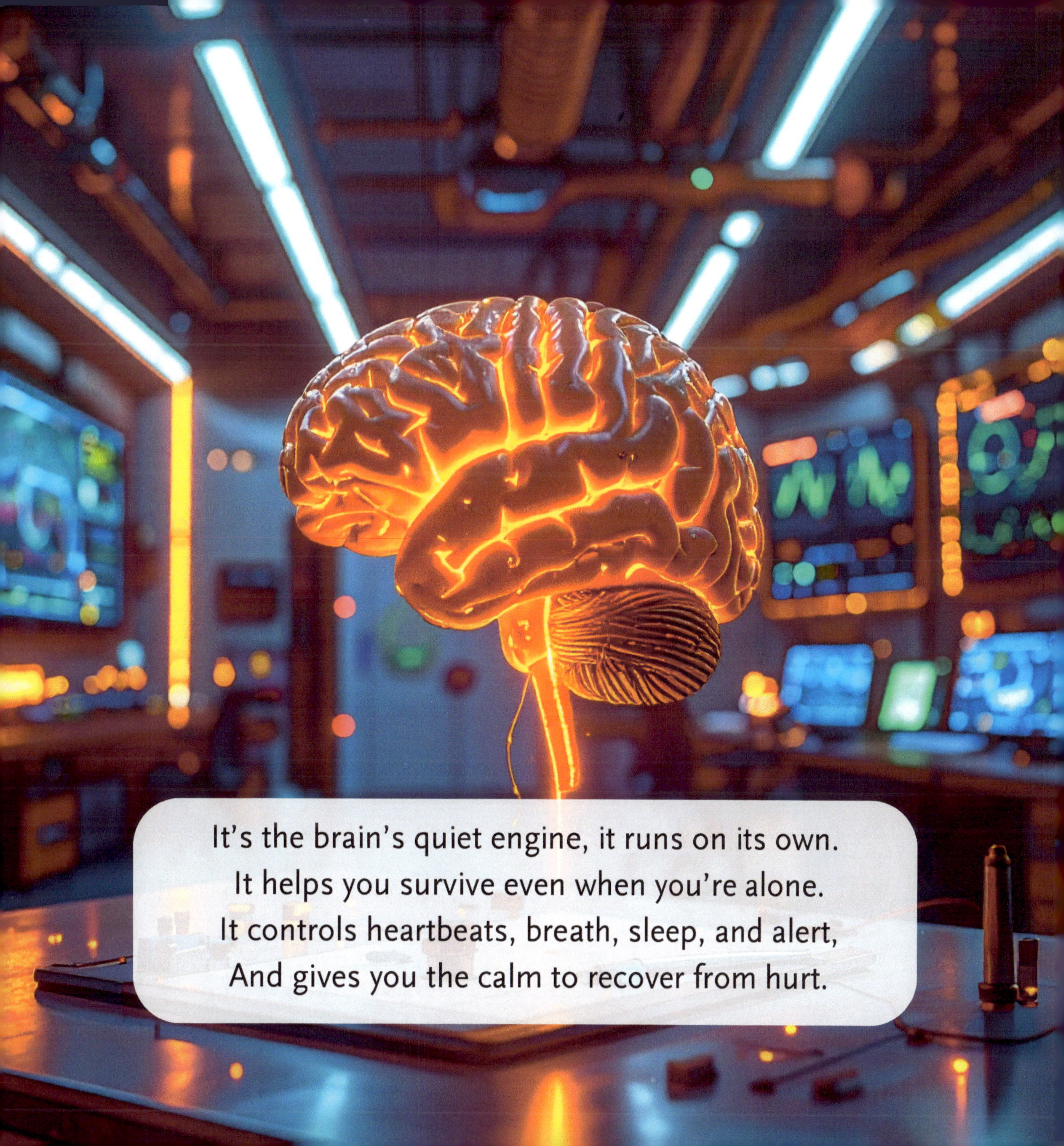

It's the brain's quiet engine, it runs on its own.
It helps you survive even when you're alone.
It controls heartbeats, breath, sleep, and alert,
And gives you the calm to recover from hurt.

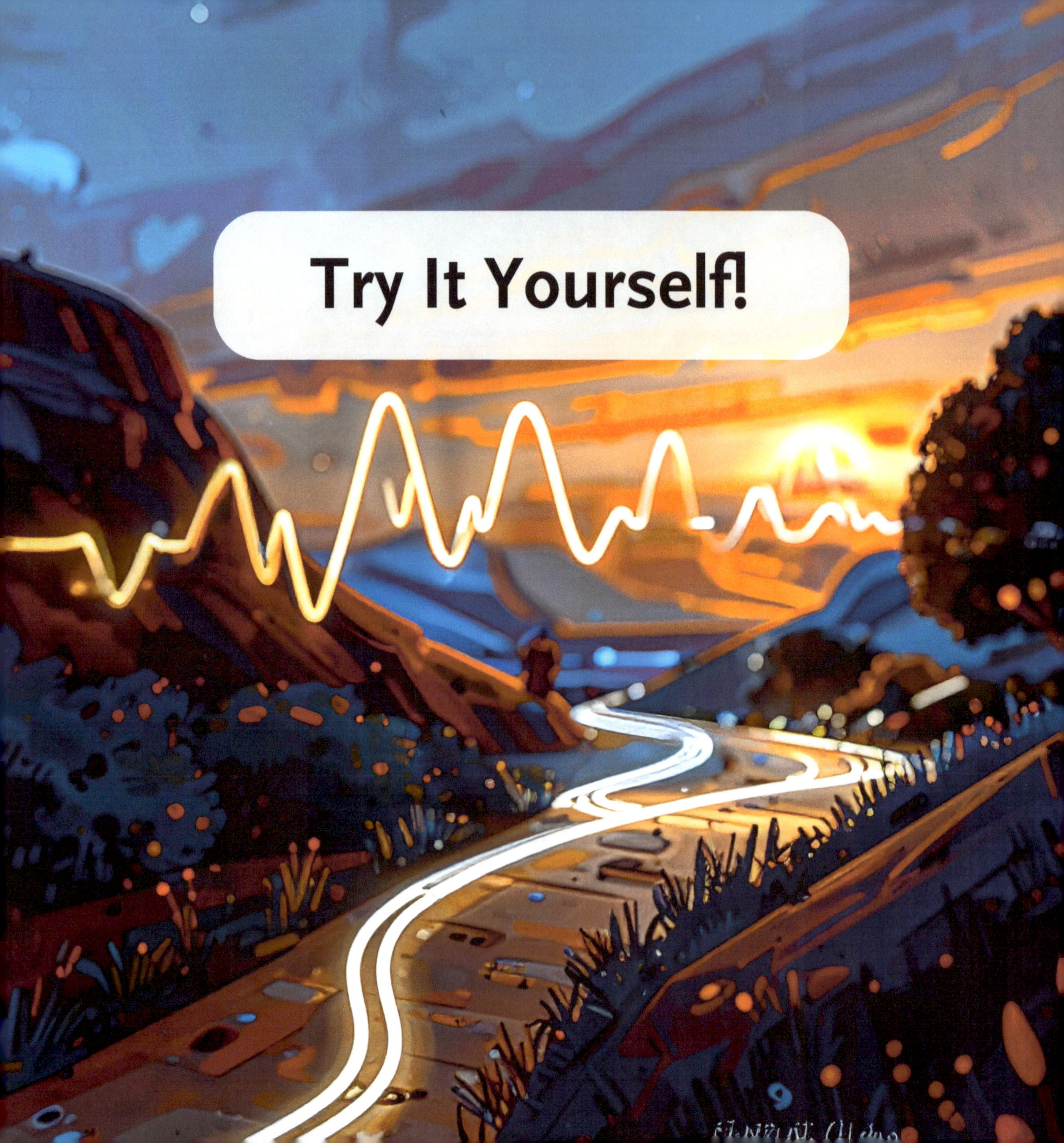
Try It Yourself!

Lie on the floor and feel your breath flow.
Count to four slow, then let the air go.
Try sips of cool water and notice each toe,
And thank your brainstem for the calm it can show.

Congratulations!

You've met Boiler Room Benny, your backstage brain,
The one who keeps going through sunshine and rain.
With every breath, you're never alone,
Because Benny is steady, deep in your zone.

Thank You to My Readers

Thank you for reading this book.
Whether you're a parent, a teacher, a therapist, or a brilliant young brain explorer,
your interest in understanding how the mind works
means the world.

Sharing these books
means spreading compassion and curiosity.

You are the reason
these stories come alive.
Thank you for believing in the power of playful learning.

Stay curious. Stay kind.
Stay brainy!

Brainstem Benny & The Automatic Adventure

Teacher & Parent Guide

💡 What This Story Teaches:

• Brain part: Brainstem Benny (automatic survival systems)

• Core idea: Some brain jobs happen quietly, all the time, without us thinking

• Skills supported: body awareness, calming, pacing, recovery

• Big message: Before learning, thinking, or focusing; our bodies need to feel safe and steady

🧠 Key Concepts for Grown-Ups

Brainstem Benny represents the brainstem, the part of the brain that keeps the body alive and regulated. It controls breathing, heart rate, alertness, and sleep-wake cycles, keeping these automatic functions in balance. When children are overtired, overstimulated, hungry, sick, anxious, or overwhelmed, this system works harder. Higher-level thinking (focus, memory, decision-making) becomes difficult until the body settles again.

This story helps children understand that:

- Feeling dizzy, tired, or "off" is not a failure
- Slowing down is a tool, not a punishment
- Calm begins in the body

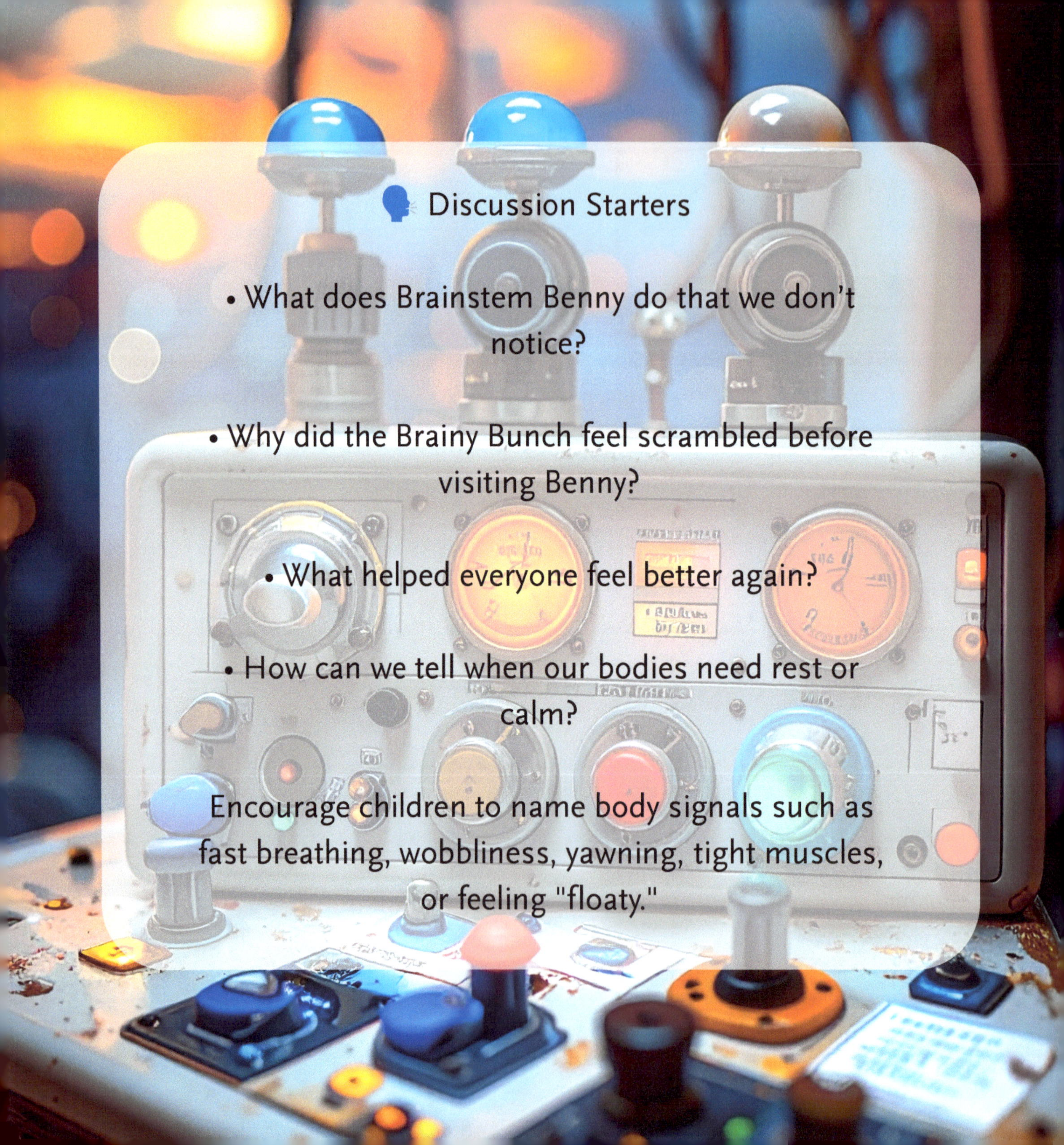

🗣️ Discussion Starters

- What does Brainstem Benny do that we don't notice?
- Why did the Brainy Bunch feel scrambled before visiting Benny?
- What helped everyone feel better again?
- How can we tell when our bodies need rest or calm?

Encourage children to name body signals such as fast breathing, wobbliness, yawning, tight muscles, or feeling "floaty."

Try This Together:

1. Breathing Reset

Practice breathing in through the nose for 4 counts and out through the mouth for 4 counts. Repeat 3 times.

2. Body Check-In

Ask: What do you notice in your chest, belly, hands, feet, or head?

3. Boiler Room Break

Create a short pause during the day for water, stretching, or quiet sitting; no problem-solving required.

4. Name the Calm

When a child settles, gently say: "That's your brainstem helping."

🎯 Bonus Challenge

Build Benny's Boiler Room

Invite children to draw or build a cozy "boiler room" using paper, blocks, or craft supplies.

Ask:

- What helps Benny keep the brain running smoothly?
- What should be inside the boiler room to help the body feel calm?

Optional labels:

- Breath Switch
- Water Station
- Stretch Zone
- Rest Button

Playful Extension:

The Brainstem Game

Call out scenarios and let kids choose a calm response:

- You feel dizzy → breathe slowly
- You feel tired → stretch or rest
- You feel jumpy → drink water and pause

Make it silly. Add movement. Let kids take turns being "Benny."

🛠️ Benny's Calm Toolbox:

This optional toolbox helps adults support children when their bodies feel overwhelmed, tired, or off-balance. These tools work best before problem-solving or teaching.

Use one or two at a time... simple is powerful.

• Breathing:

Inhale through the nose for 4,
exhale through the mouth for 4.
Repeat 2-3 times.

• Water Reset:

Offer a sip of cool water.
Swallowing helps the body settle.

• Humming or Soft Sounds:

Humming, gentle music, or quiet singing
can calm the nervous system.

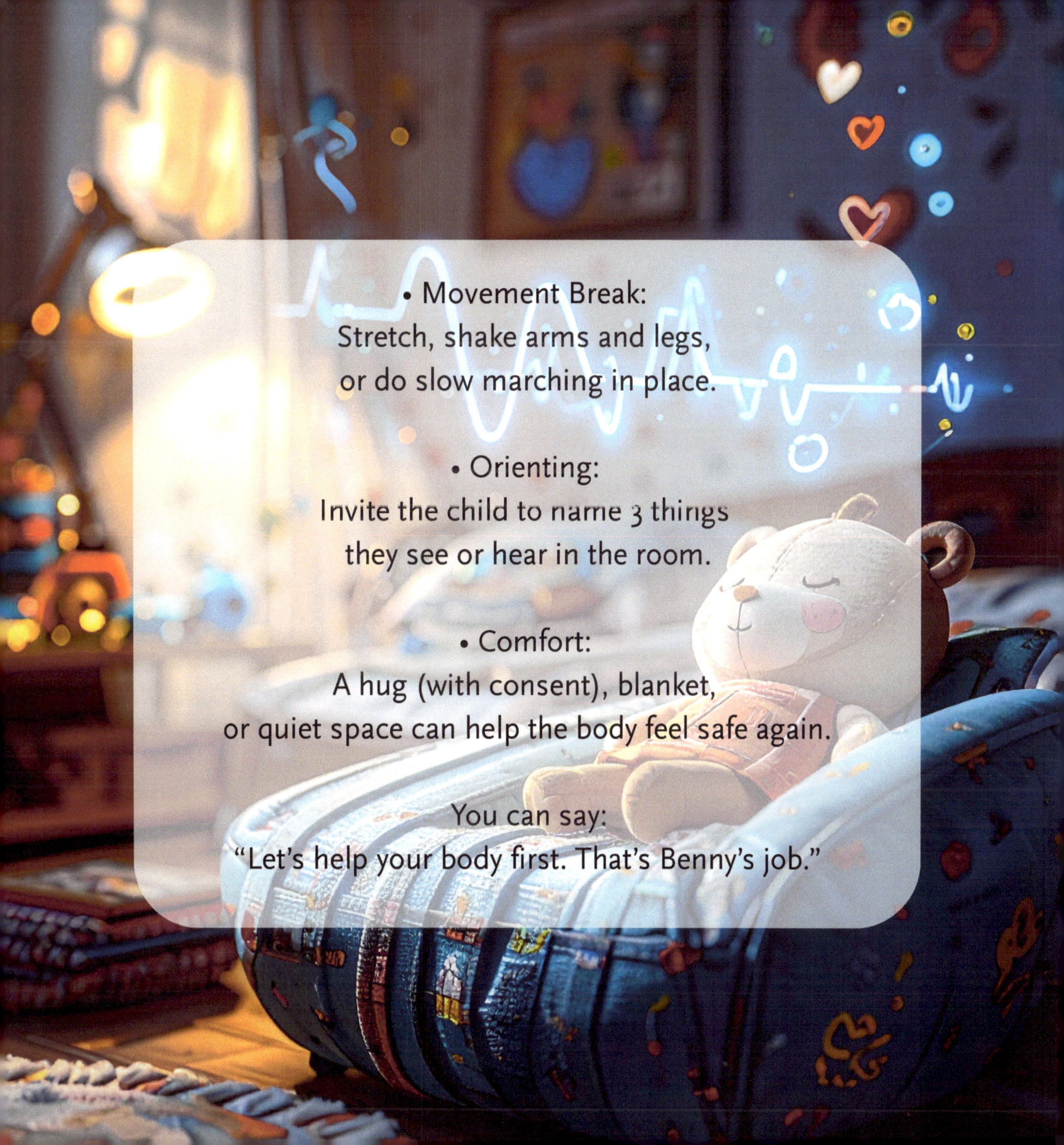

• Movement Break:
Stretch, shake arms and legs,
or do slow marching in place.

• Orienting:
Invite the child to name 3 things
they see or hear in the room.

• Comfort:
A hug (with consent), blanket,
or quiet space can help the body feel safe again.

You can say:
“Let’s help your body first. That’s Benny’s job.”

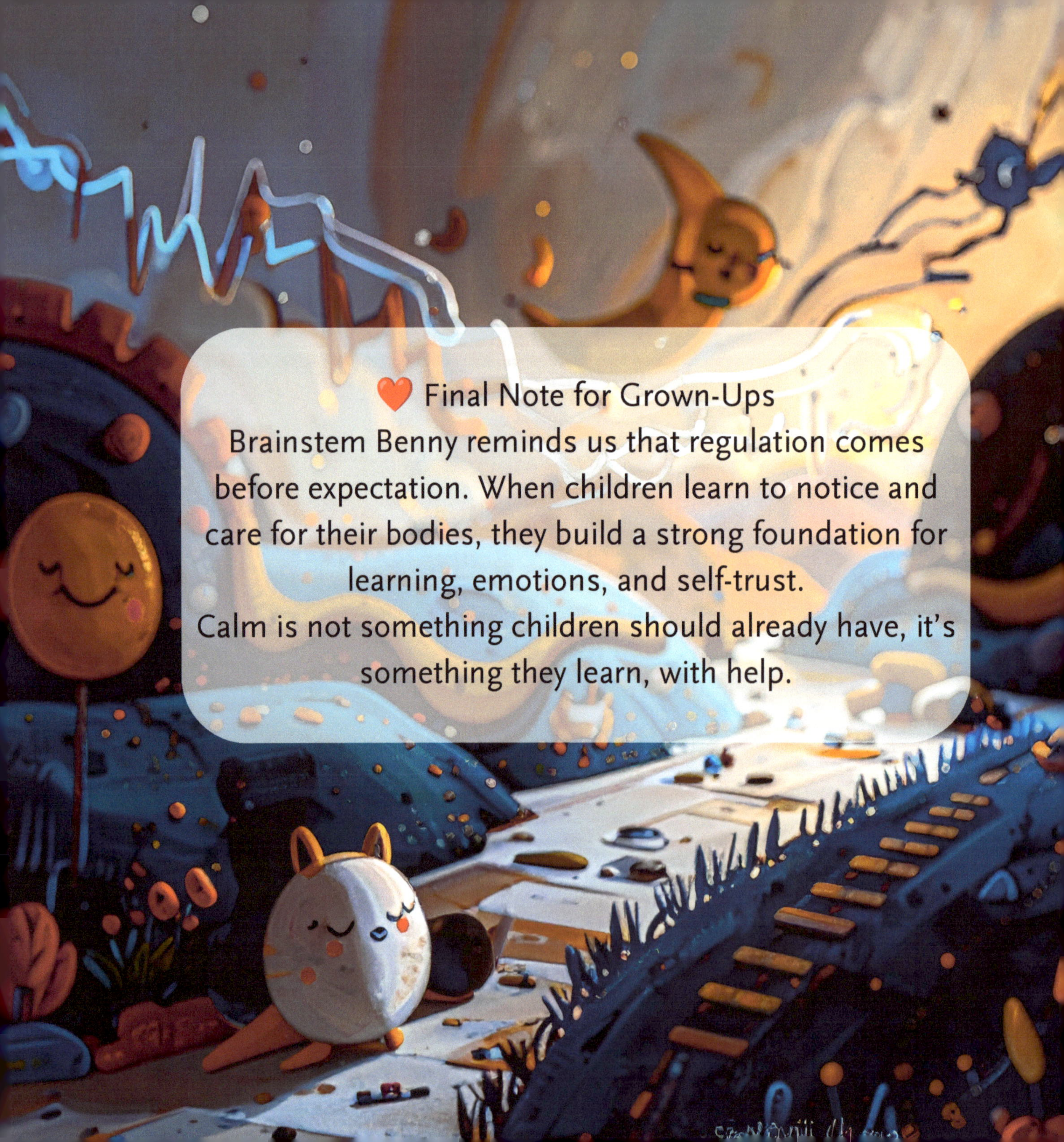

❤️ Final Note for Grown-Ups

Brainstem Benny reminds us that regulation comes before expectation. When children learn to notice and care for their bodies, they build a strong foundation for learning, emotions, and self-trust.

Calm is not something children should already have, it's something they learn, with help.

Resources & GROWTH

Resources:

(Brainstem + automatic survival functions like breathing, heartbeat, blinking, swallowing)

- NINDS (NIH). Brain Basics: Know Your Brain (brainstem overview).
- NIDA (NIH). Brain Power! Challenge: Module 1- Introduction to the Brain and Nervous System. (PDF).
- Cleveland Clinic. Brain: parts and function (brainstem/autonomic functions).
- NINDS (NIH). Brain Basics: Know Your Brain / Brain Basics resources. (2024-2025).
- Cleveland Clinic. Brain: parts and function (overview pages). (2025 update).
- CASEL. CASEL's SEL Framework (self-awareness, self-management, social awareness, relationship skills, responsible decision-making). (2020, updated).
- CDC. Child Development: Positive Parenting Tips and developmental guidance by age. (2024-2025).
- MedlinePlus (U.S. National Library of Medicine)
- Encyclopedia Britannica

Growth:

(If you or someone you know wants to strengthen calming, body regulation, or back-to-steady routines)

- Harvard Center on the Developing Child
  - Look for: self-regulation, stress response
- CASEL (Collaborative for Academic, Social, and Emotional Learning)
  - Look for: self-management
- American Academy of Pediatrics (HealthyChildren.org)
  - Search: calming strategies, sleep routines

Acknowledgments:

To my students: past, present, and future:
You taught me more than any textbook ever could. By watching how you explored, questioned, struggled, and discovered, I learned how minds truly grow. Every "aha" moment in these pages began with you.

To the parents, teachers, and guides walking beside children every day:
Thank you for listening, for advocating when things feel complicated, and for believing that understanding a child's inner world truly matters. Your patience and care shape futures in ways you may never fully see.

To my husband, my best friend, and my constant partner in learning:
Thank you for walking beside me, believing in this work, and holding space for creativity to exist. Sharing curiosity, growth, and creation with you has been one of the greatest honors of my life.

To the great thinkers who shaped how I see the world: including Dr. Seuss, Albert Einstein, Leonardo da Vinci, Stephen Hawking, Ludwig van Beethoven, just to name a few, and my English teacher, Mrs. Godfrey:
Your curiosity, courage, and imagination reminded me that learning is limitless, creativity is powerful, and questions matter. Mrs. Godfrey, your encouragement, belief in me, and love of words sparked my writing journey. If not for you, I may never have picked up a pen.

And to my publisher, NextGen Learners™
It was by fate that we met, and by inspiration that we now move forward. Together, we build something meaningful. Together, we leave this legacy.

Lastly, to my own brain:
Complex, resilient, ever-adapting, and endlessly learning. Thank you for carrying stories, holding wonder, and reminding me that growth is always possible.

Maisel McLaula is the author of the Think Like a Brain!™ children's book collection, a story-first series that blends neuroscience, imagination, and emotional safety for young readers.

Drawing from decades of experience supporting children and families, her work centers on a simple but powerful belief: children deserve to understand and accept who they are, including how their brains work and why they make the choices they do.

Through playful characters and gentle metaphors, Maisel's stories help children discover the "why" behind their thoughts, feelings, and actions, turning confusion into clarity and self-judgment into self-compassion. By understanding their inner systems, children gain the power to respond with kindness toward themselves and to recognize those same patterns in others.

Maisel believes that when children feel safe inside their own minds, empathy grows naturally. The Think Like a Brain!™ collection offers a foundation for confidence, understanding, and care that grows with the child.

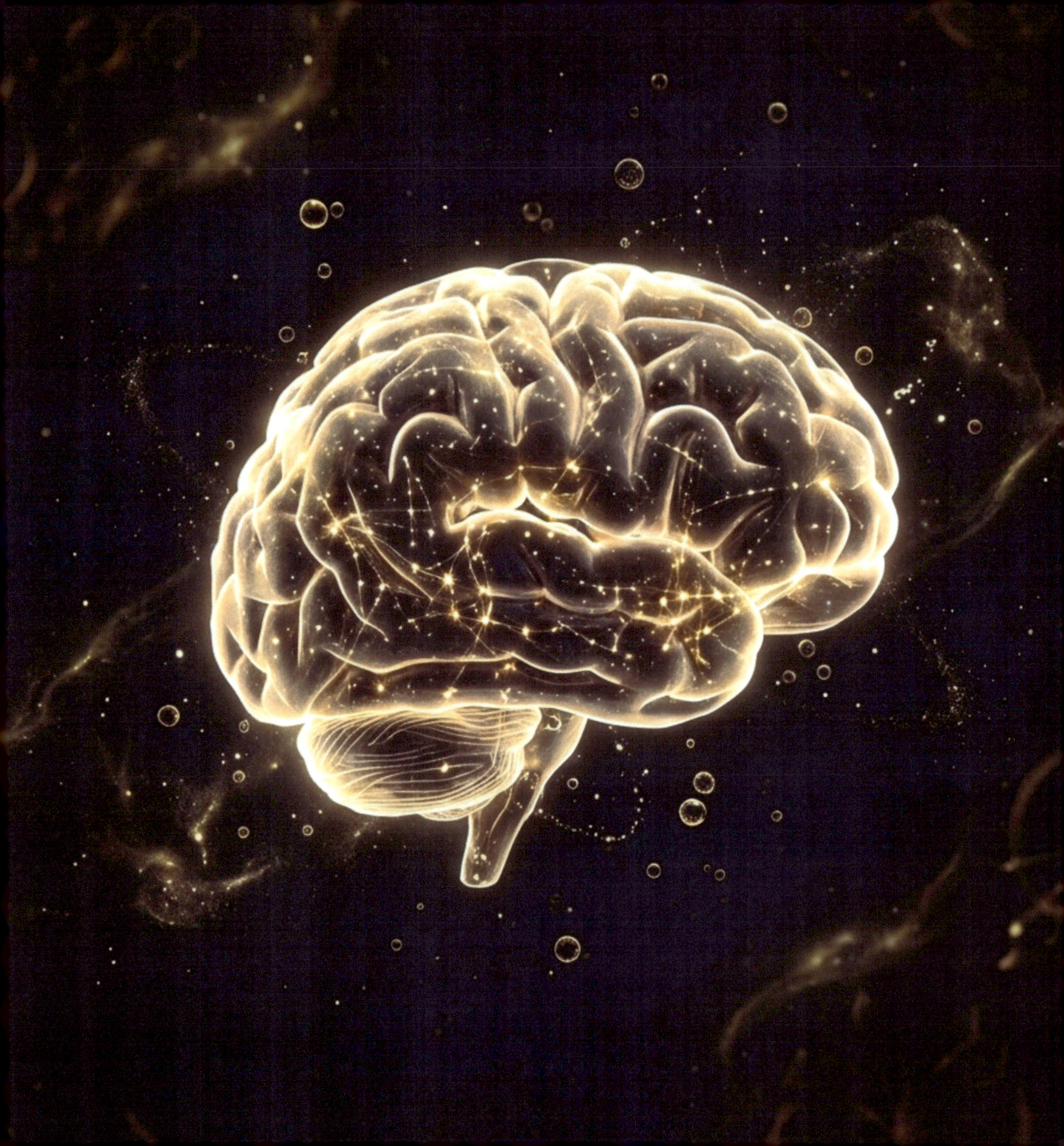

ABOUT THE PUBLISHER

NextGen Learners LLC is an educational publishing company dedicated to helping children understand how they think, feel, learn, and grow.

Through imaginative storytelling, neuroscience-informed concepts, and real-world application, NextGen Learners™ creates books and resources that support emotional development, critical thinking, self-awareness, and confidence in young minds.

Our work is designed not only for children, but for the parents, educators, caregivers, and professionals who walk beside them. Each book serves as a bridge, helping adults better understand the child in front of them, while helping children better understand themselves.

By creating shared language and meaningful connection points, our stories foster deeper relationships, reduce misunderstanding, and encourage acceptance, both within oneself and toward others.

We believe every child's mind is uniquely wired, and that understanding those differences is the foundation for compassion, growth, and lifelong learning.
With multiple collections designed to support cognitive, emotional, and developmental growth, NextGen Learners provides tools that children can carry with them, and tools that adults can use to guide them with clarity, patience, and purpose.

MORE TO EXPLORE

The Brainy Bunch is just the beginning.
More stories, tools, and brain-friendly resources are on the way! Designed to help children, parents, and educators better understand how the brain works in everyday life.

FREE PRINTABLES

We've created printable pages to go along with this book, including:

-Reading Guides
-Brain helper visuals
-Character Cards

These are available for free as part of our growing resource library.

Our full website experience is built to include:

-Downloadable activity packs
-Parent & teacher guides
-Expanded learning tools
-Additional collections and topics

STAY CONNECTED

We love hearing how children connect with these stories!

If there is a topic, feeling, or challenge you would like to see explored in future books, we would love to hear from you!

Contact us: Inspire@NextGenLearners.com
NextGenLearners.com

Your ideas help shape future stories and support the growth of this collection.

This collection grows with you. New resources and stories are added regularly.

Maisel McLaula loves hearing from the tiny humans who read her books.

Did a story make you giggle, spark a big idea, or inspire you to draw your very own character? She would be delighted to see it! Send your letters and artwork to:

NextGen Learners
Attn: Maisel McLaula
8666 Beechmont Ave #1063
Cincinnati, OH 45255

Every letter is read with a smile, and your artwork just might make her whole day.

OPEN BRAIN

www.ingramcontent.com/pod-product-compliance
Lightning Source LLC
Chambersburg PA
CBHW041322210926
51847CB00002B/24
9781972860014